CRASH COURSE BIOLOGY: A STUDY GUIDE OF WORKSHEETS FOR BIOLOGY

BY ROGER MORANTE

Library of Congress Cataloging-in-Publication Date is available.

ISBN-13: 978-1-7322125-6-5

Writer: Roger Morante
Cover Design: Artwork purchased from 99Designs.com.
Cover Artist: Edzel Davo Antiquina (Righteous.Juan)
Editor: Roger Morante
Copy Editor: Erica Brown
Back Cover Photo: Liesl Morante
Publisher Logo: Isabella Morante

To contact the publisher, send an email to the address below;
holden713@gmail.com

Additional copies may be purchased on Amazon.com or by contacting the author.

Printed in the United States of America

First printing January 2020

TABLE OF CONTENTS

28) THE DIGESTIVE SYSTEM - CRASH COURSE BIOLOGY #28
29) THE EXCRETORY SYSTEM: FROM YOUR HEART TO THE TOILET – CRASH COURSE BIOLOGY #29
30) THE SKELETAL SYSTEM: IT'S ALIVE! – CRASH COURSE BIOLOGY #30
31) BIG GUNS: THE MUSCULAR SYSTEM – CRASH COURSE BIOLOGY #31
32) YOUR IMMUNE SYSTEM: NATURAL BORN KILLER – CRASH COURSE BIOLOGY #32
33) GREAT GLANDS – YOUR ENDOCRINE SYSTEM: CRASH COURSE BIOLOGY #33
34) THE REPRODUCTIVE SYSTEM: HOW GONADS GO – CRASH COURSE BIOLOGY #34
35) OLD & ODD: ARCHAEA, BACTERIA & PROTISTS - CRASH COURSE BIOLOGY #35
36) THE SEX LIVES OF NONVASCULAR PLANTS: ALTERNATION OF GENERATIONS – CRASH COURSE BIOLOGY #36
37) VASCULAR PLANTS = WINNING! - CRASH COURSE BIOLOGY #37
38) THE PLANTS & THE BEES: PLANT REPRODUCTION – CRASH COURSE BIOLOGY #38
39) FUNGI: DEATH BECOMES THEM – CRASH COURSE BIOLOGY #39
40) ECOLOGY – RULES FOR LIVING ON EARTH: CRASH COURSE BIOLOGY #40

Name___________________
Period__________________
Date___________________

Carbon… SO SIMPLE: Crash Course Biology #1

1) Explain why the **carbon** atom has the ability to form itself into weird rings, sheets, spirals, double, and triple bonds.

____
____
____
____
____
____

2) Analyze the reasons why **carbon** is the core **atom** of complicated structures that make up living things.

____
____
____
____

3) Why do YOU think that many scientists have difficulty conceiving of a life form that is NOT made out of **carbon**?

____
____
____
____
____
____

4) Evaluate how the **covalent** structure of **carbon** allows it to have empty spaces in its **shells** so that it can pair up with other **atoms**.

____
____
____
____
____
____

5) Assess the ways in which the compound **methane** utilizes **covalent bonds** between **carbon** and **hydrogen** in order to form itself into a **molecule**.

____
____
____
____
____
____

6) Explain the reasons why YOU think that **Lewis Dot Structures**, made up by Gilbert Lewis (1875-1946), are so useful in understanding the **octet rule** in chemistry.

____

____

____

____

____

7) Explain how water (H_2O) is comprised using the **octet rule**.

____

____

____

____

____

8) Explain how **nitrogen** can form into the compound **ammonia** (NH_3), and then be part of an **amino group** and then become an **amino acid**.

____

____

____

____

____

____

9) Explain the reasons why the **molecule** H_20 forms a **polar covalent bond**.

____

____

____

____

____

10) Evaluate the problem that the **ionic compound** salt, or **sodium chloride** (NaCl), has with sharing **electrons** preferring to exist as an **ionic bond**.

____

____

____

____

____

____

11) Analyze how the **hydrogen bond** in a molecule of H_2O works to create the **polar covalent bond** in that **molecule** of water.

____

____

____

____

____

____

Name__________________
Period_________________
Date__________________

Water – Liquid Awesome: Crash Course Biology #2

1) Explain the reasons why the **covalent bonds** found inside of **water** are so important in creating **biological molecules** for life.

____
____
____
____
____
____

2) Analyze the reasons why the **polarity** of **hydrogen bonds** inside of water causes high **cohesion** inside of the molecule.

____
____
____
____
____
____

3) Clarify how the **surface tension** of water causes water to bead up.

____
____
____
____
____

4) Formulate a hypothesis surrounding how **adhesion** causes water to stick to glass instead of beading up.

____
____
____
____

5) Why do YOU think that it is cool that water can defy **gravity**? *If the aforementioned question rings true, and YOU had all the resources in the world, imagine up an invention YOU would design to change the world.*

____
____
____
____
____
____
____

6) Investigate and report upon the **polarity** of sugar and salt, and then explain why their molecules are **hydrophilic** when interacting with the **cohesive forces** of **water**.

7) Explain what happens to a **hydrophobic** molecule which cannot break the **cohesive forces** of water even though water has many times been coined the universal **solvent**.

8) Clarify the main reason why scientist Henry Cavendish (1731-1810) didn't get credit for discovering **Richter's Law**, **Ohm's Law**, and **Coulomb's Law**.

9) Hypothesize as to why **hydrogen bonds** cause solid ice cubes to float in liquid water.

10) Connect how the **heat capacity** of water allows water to hold onto heat by analyzing what would happen to all life on the land and the oceans of the earth during a significant **global warming** event. *(*significant: temperature rise of 2°C=3.6°F)*

Name____________________
Period__________________
Date____________________

Biological Molecules – You Are What You Eat: Crash Course Biology #3

1) Evaluate the reasons why **biological molecules** are essential and necessary for every living thing on earth. **Hint – imagine if there were no biological molecules on earth.*

____
____
____
____
____
____

2) Assess the factors which led physician William Prout (1785-1850) to be the first to discover that human **stomachs** contained **hydrochloric acid** as well as the factors that led the discovery of the chemical composition of **urine**.

____
____
____
____

3) Explain how **carbohydrates** are broken down in the human body.

a) **Monosaccharides -glucose** and **fructose**

____
____
____
____

b) **Disaccharides -sucrose**

____
____
____

c) **Polysaccharides – cellulose**

____
____
____
____

4) Briefly analyze the reasons why human beings can eat bread but not eat grass.

____
____
____
____

5) Briefly explain how **glycogen** plays a role in storing energy inside of a person's muscles.

____
____

6) Explain the reasons why **lipids** have difficulty dissolving in water.

____

____

____

7) Compare the two types of fats, **glycerol** and **fatty acids**, and then explain how they come together to form the **triglycerides** found in butter, peanut butter, and oils.

____

____

____

____

____

8) Clarify the differences inside of the placement of **carbon bonds** between **saturated** and **unsaturated fatty acids** found inside of a jar of peanut butter.

____

____

____

____

9) Evaluate the effectiveness of hydrophobic and hydrophilic **phospholipids** in making up **cell membrane** walls.

____

____

____

____

10) Briefly explain the advantages of the **four-carbon ring** structure of the **steroid** family by analyzing **cholesterol**, the most abundant **steroid** in the body.

____

____

____

____

11) Analyze the reasons why **proteins** are the most complicated and awesome **compounds** in the human body. **Enzymes-antibodies-protein hormones*

____

____

____

____

____

____

12) Clarify how an egg can supply a person with digestible **nitrogen**.

____

____

____

____

____

Name__________________
Period_________________
Date__________________

Eukaryopolis – The City of Animal Cells: Crash Course Biology 4

1) Compare the advantages of **cell membranes** between those inside of plants and those inside of animals.

____
____
____
____

2) Critique scientist Robert Hooke's (1635-1703) analysis of what he thought was going on inside of a **eukaryotic cell** by comparing it to what scientists now know today.

____
____
____
____
____

3) Analyze the differences between **cilia** and **flagellum** inside of **eukaryotic cell**. *Include how **microtubules** play a significant role inside of the cell.*

____
____
____
____

4) Explain the function of **selective permeability** in a **cell membrane** by relating it to fascist border police.

____
____
____

5) Relate how the **cytoplasm** is like a swamp inside of the cell city of a **eukaryotic cell**. *Include the role **centrosomes** play inside of this swampy city.*

____
____
____
____
____

6) Clarify the differences between the **cytoplasm** in a cell and the **nucleoplasm** inside of the **nucleus** of a **eukaryotic cell**.

____
____
____
____
____

7) Evaluate the effectiveness of the **Endoplasmic Reticulum** (**ER**), aka the highway system, inside of a **eukaryotic cell**.

___

___

___

___

8) Briefly explain the following functions of the **Smooth ER** inside of a **eukaryotic cell**:
   a) Factory – Warehouse

   ___

   ___

   b) Enzymes >>Lipids

   ___

   ___

   c) Cell Detox

   ___

   ___

   d) Stores Ions

   ___

   ___

9) Briefly explain the function of the **Rough ER** inside of a **eukaryotic cell**.

- Ribosomes >>Amino Acids >>polypeptides

___

___

___

___

___

10) Briefly explain the function of the **Golgi Apparatus** inside of a **eukaryotic cell**.
   A) Protein processing & packaging

   ___

   ___

   B) Sends products out

   ___

   ___

11) Analyze the function of **Golgi Bodies** inside of a **eukaryotic cell**.
   A) Golgi Apparatus Layers

   ___

   ___

   B) Proteins >> Hormones

   ___

   ___

   C) Proteins + Carbs = Molecules

   ___

   ___

Name____________________
Period__________________
Date____________________

In Da Club – Membranes & Transport: Crash Course Biology #5

1) Describe how cells have **selectively permeable cell membranes**, and are similar to how nightclubs pick and choose who can enter and who needs to wait outside in line.

____
____
____
____

2) Explain how different materials move in and out of **cell membranes** using **passive transport** through a process called **diffusion**.

____
____
____
____
____

3) Clarify how **active transport** requires the usage of cellular energy such as **adenosine triphosphate** (ATP) in order to achieve movement.

____
____
____
____

4) Analyze the reasons why water diffuses in and out of a **cell membrane** through a process called **osmosis**.

____
____
____
____
____

5) Differentiate between **hypertonic** and **hypotonic** solutions when analyzing why water wishes to move across its concentration gradient to become an **isotonic** solution.

____
____
____
____
____

6) Describe the function of the **kidneys** in regulating **blood** in a body to keep it **isotonic**.

____
____
____
____

7) Analyze how **channel proteins** inside of the **phospholipid bilayer** of a **cell membrane** allow water and ions into a cell without using any energy.

8) Explain how **ATP** moves something in the opposite direction from a **low concentration** gradient to a **high concentration** gradient in a cell.

9) Compare how **ATP** is similar to using a credit card in the **sodium potassium pump** inside of muscle cells and brain cells.

10) Analyze how proteins get the energy to power their **sodium potassium pump** and transport ions both inside and outside of a **cell membrane** of a **nerve cell** so that a person can have the **electrical chemical energy** needed to for sensation.

11) Explain the way a person's brain releases **neurotransmitters** through the **vesicular transport** process called **exocytosis.**

12) Explain the way a person's brain accepts **neurotransmitters** through the **vesicular transport** process called **endocytosis.**

Name___________________
Period_________________
Date___________________

Plant Cells: Crash Course Biology #6

1) Clarify how the existence of the **Carboniferous Period** (359-299 million years ago) - along with **scale trees** - is important to the car you drive or want to drive today. ** Hint - Think about the way the world would be today without the coal derived from the **Carboniferous Period**.*

____
____
____
____
____
____

2) Evaluate the reasons why **eukaryotic cells** (separately enclosed **nucleus** and **organelles**) are more advanced then **prokaryotic cells** (**bacteria**).

____
____
____
____
____
____

3) Explain the basic differences between **plant cells** and **animal cells**.
   a) **Plant cells (lignin, cellulose, cell walls)**

   ____
   ____
   ____
   ____
   ____

   b) **Animal cells** (flexible **membrane**)

   ____
   ____
   ____
   ____
   ____

4) Differentiate between the reasons why people can't eat wood like a beaver or grass like a cow or a goat. *(Include the role **bacteria** plays in the **digestion** process)*

____
____
____
____

5) Analyze the reasons why **cellulose** and **lignin** are very useful to humans even though humans cannot digest **cellulose** or **lignin**.

6) Explain the reasons why **photosynthesis** is so important to the production of food inside of a **plant cell**. *(Include the roll **plastids** play in making and storing compounds)*

7) Clarify how **chloroplasts** inside of **plant cells** make food and oxygen for human beings.

8) Clarify the function of a **vacuole** inside of a **plant cell**. *(Explain **turgor pressure** in your answer.)*

9) Analyze the reasons why celery gets soft after being inside of a refrigerator for too long. *Be sure to clarify the role that **turgor pressure** plays in this process.*

10) Point out and explain the basics of **plant cell anatomy** that make up a plant.

Name____________________
Period__________________
Date____________________

ATP & Respiration: Crash Course Biology #7

1) Chemically explain how YOU derive energy from food (**glucose**) to power **cellular respiration** so that YOU can work out at the gym or play sports.

2) Examine and report as to how the currency of biological energy, **ATP** (**Adenosine Triphosphate**), allows cells to grow, move, and create electrical impulses by shooting off a **phosphate molecule** to create **ADP** (**Adenosine Diphosphate**).

3) Briefly analyze how **hydrolysis** adds a water molecule to a substance after one of the three **ATP phosphate molecules** is ejected to form the **ADP compound**.

4) Explain how **glycolysis**, or the breaking down of **glucose** into **pyruvate acids**, can help a person who is lifting weights to make **lactic acid**.

5) Cite evidence as to how the **Krebs Cycle** involves the **mitochondria** of a cell in order to generate energy through **aerobic respiration**. *Include the chemical breakdown of **pyruvate molecules, glucose,** and **enzymes** in your answer.*

____

____

____

____

____

____

____

____

6) Analyze how **enzymes** help to join the **acetyl coA** and **Oxaloacetic Acid** in order to form **Citric Acid**.

____

____

____

____

____

7) Briefly explain how **CO_2** (**carbon dioxide**) exhaled by humans is connected to the **Krebs Cycle**.

____

____

____

____

8) Clarify how the battery-like enzymes **NAD+** and **FAD** store energy after being made during the **Krebs Cycle** for later use by the cell in the **electron transport chain**.

____

____

____

____

____

____

____

9) Evaluate the effectiveness of the **electron transport chain** in working as a pump to channel **proteins** across the **inner membrane** of the **mitochondria** through the process called **ATP synthase**.

____

____

____

____

____

____

Name____________________
Period__________________
Date____________________

Photosynthesis: Crash Course Biology #8

1) Explain how water, carbon dioxide, and sunlight lead to the two different reactions of **photosynthesis**, **light-dependent reactions** and **light-independent reactions**, and is basically just **respiration** in reverse.

____
____
____
____
____
____
____

2) Analyze the structure of **chlorophyll**, and then explain how a bunch of stacked **thylakoids** capture sunlight inside of the **chloroplast organelle** using **light-dependent reactions**.

____
____
____
____
____
____
____
____

3) Briefly explain how the process of **photoexcitation** in **PSII** (**Photosystem II**) converts **electrons** into something that the plant can use for food and energy.

____
____
____
____
____
____

4) Investigate and report upon how the **electron transport chain** causes **electrons** to lose their energy in a series of reactions in order to capture the energy people need to breathe.

____
____
____
____
____
____

5) Explain both the process and reasons why the **Cytochrome Complex** reacts with **PSII** and **PSI** (**Photosystem I**) in order to fill **thylakoids** up with **protons**.

6) Briefly explain how **PSI** makes different products than **PSII**. *Be sure to include the role **NADP+** plays in linking electrons and hydrogen atoms to form **NADPHs**.*

7) Analyze stage 2 of the **Calvin Cycle**, aka **light-independent reactions**, which occur inside of a **chloroplast** during **photosynthesis**.

8) Connect how plants converted **carbon dioxide** ($CO2$) in the early atmosphere to **oxygen** (O_2), and in order to survive rising levels of oxygen, the plants used an **enzyme** in **chloroplasts** called **rubisco** during the **Calvin Cycle** to stay alive.

9) Clarify how the reduction stage of the **Calvin cycle** converts 3-PGA into a three-carbon sugar and then show how **NADPH** is reduced to form glyceraldehyde-3-phosphate.

Name___________________
Period_________________
Date___________________

Heredity: Crash Course Biology #9

1) Clarify what is meant by the term **heredity**.

2) Compare the differences in thought about **heredity** between Aristotle (384-322 BCE) and the **classical genetics** model of Gregor Mendel (1822-1884).

3) Briefly analyze the basic differences between **genes** and **chromosomes**.

4) Clarify the relationship between **pleiotropic genes** and **polygenic traits**.

5) Investigate the **amino acids** glycine and arginine, and report on their placement on **human chromosome 16** of the **human genome**. *Then explain how the **gene**, aka **allele**, on that **gene** determines whether or not your ear wax will be wet or dry.*

6) Explain why **heredity** and the **diploid** nature of **somatic cells** affects YOUR earwax.

7) Analyze the **phenotype** of YOUR earwax by determining which types of **alleles**, recessive or dominant, make up YOUR earwax. *If you can ask your parents if they have wet or dry earwax, please do so. If not, take a guess based upon this cross-breeding experiment using a* ***Punnett square*** *as to what types of earwax YOU think YOU have based upon the composition of YOUR own earwax.*
Heterozygous = (Ff)
Homozygous dominant = (FF)
Homozygous recessive = (ff)

________________ = (%)
________________ = (%)

8) Briefly explain the correlation between wet ear wax and body odor. *Include how this is an example of a* ***pleiotropic gene****.*

9) Identify and analyze the differences between **chromosome 23** in males and females that determine our sex, aka **sex-linked inheritance**.

Name___________________
Period_________________
Date___________________

DNA Structure and Replication: Crash Course Biology #10

1) Clarify the reasons human **DNA** (**deoxyribonucleic acid**), a type of **nucleic acid,** stores genetic instructions inside of a cell.

____

____

____

____

2) Explain how **polymers**, specifically **nucleotides**, comprise the structure of **DNA**.

____

____

____

____

3) Analyze the genetic coding of a **nucleotide** by clarifying the function of its three parts: 5-carbon sugar molecule, a phosphate group, and one of four nitrogen bases which makes up the structure **double helix**.

____

____

____

____

____

____

____

____

4) Assess how a **double helix** is made by clarifying why just any placement of a **nitrogenous base** would not work in its construction. *Explain the* ***base pair*** *placement of* ***adenine*** *(A),* ***thymine*** *(T),* ***guanine*** *(G), and* ***cytosine***(*C) in your answer.*

____

____

____

____

____

____

5) Explain how the **base sequence** of a **nitrogenous base** works to create YOU.

____

____

____

____

____

____

6) Explain how it is possible to use one **base sequence** in **DNA** to predict what its matching sequence will happen to be.

____

____

____

____

7) Clarify the three major differences as to why the construction of base sequences in **ribonucleic acid** (RNA) is different than that of **DNA**.

a) ____

____

b) ____

____

c) ____

____

8) Evaluate the importance that the **X-Ray diffraction** studies of Rosalind Franklin (1920-1958) played in discovering the **helical structure** of the **double helix**.

____

____

____

____

____

9) Analyze how the enzyme **helicase** unzips a **double helix** in order to duplicate a cell in a process called **replication**.

____

____

____

____

____

____

____

____

10) Assess the steps **DNA polymerase** takes to build a new **DNA chain** and include why it needs an **RNA primer**, aka the **Okazaki fragments**, in order to start building this new **DNA chain**.

____

____

____

____

____

____

____

____

____

Name___________________
Period_________________
Date___________________

DNA, Hot Pockets, & The Longest Word Ever: Crash Course Biology #11

1) Briefly explain how humans are made starting from the **DNA transcription** inside a cell's nucleus and then onto the process of **translation** that allow humans to build **proteins** from **enzymes** and **carbohydrates**. *Relate this to the construction of a hot pocket.*

____
____
____
____
____
____
____
____
____
____

2) Briefly explain the relationship between an RNA molecule, DNA, and a **transcription unit**.

____
____
____
____
____
____

3) Clarify the reasons why the **Tata box** helps **enzymes** figure out where to bind to either the upstream or the downstream strand of DNA.

____
____
____
____
____
____
____

4) Clarify why **uracil** appears in thymine's place as a partner to **adenine** in the **RNA polymerase** and on up to the **termination signal** instead of thymine like DNA.

____
____
____
____
____
____

5) Briefly analyze the reasons for the finishing touches that occur in the **Messenger RNA (mRNA)** package once the **5-prime cap** is copied up to the 3-prime end **poly-A tail**.

6) Explain how **snRNPs** (small nuclear ribonucleoproteins), along with other proteins, form into a large and complex **spliceosome** enabling **translation** of **mRNA** into a protein (**RNA splicing**), and then compare how it is similar to editing a video.

7) Differentiate between **introns** and **exons** making sure to explain how recycling in the nucleus occurs. *Analyze an example of a protein with base pairs that falls into this category.*

8) Explain how **translation** is a process of **gene expression** where **mRNA** is decoded in the **ribosome decoding center** and involves a blueprint of instructions with ingredients to produce a specific **amino acid chain** (aka **polypeptide**).

Name____________________
Period__________________
Date____________________

DNA, Hot Pockets, & The Longest Word Ever: Crash Course Biology #11

1) Briefly explain how humans are made starting from the **DNA transcription** inside a cell's nucleus and then onto the process of **translation** that allow humans to build **proteins** from **enzymes** and **carbohydrates**. *Relate this to the construction of a hot pocket.*

____
____
____
____
____
____
____
____
____
____

2) Briefly explain the relationship between an RNA molecule, DNA, and a **transcription unit**.

____
____
____
____
____
____

3) Clarify the reasons why the **Tata box** helps **enzymes** figure out where to bind to either the upstream or the downstream strand of DNA.

____
____
____
____
____
____
____

4) Clarify why **uracil** appears in thymine's place as a partner to **adenine** in the **RNA polymerase** and on up to the **termination signal** instead of thymine like DNA.

____
____
____
____
____
____

5) Briefly analyze the reasons for the finishing touches that occur in the **Messenger RNA** (**mRNA**) package once the **5-prime cap** is copied up to the 3-prime end **poly-A tail**.

6) Explain how **snRNPs** (small nuclear ribonucleoproteins), along with other proteins, form into a large and complex **spliceosome** enabling **translation** of **mRNA** into a protein (**RNA splicing**), and then compare how it is similar to editing a video.

7) Differentiate between **introns** and **exons** making sure to explain how recycling in the nucleus occurs. *Analyze an example of a protein with base pairs that falls into this category.*

8) Explain how **translation** is a process of **gene expression** where **mRNA** is decoded in the **ribosome decoding center** and involves a blueprint of instructions with ingredients to produce a specific **amino acid chain** (aka **polypeptide**).

Name____________________
Period___________________
Date____________________

Mitosis: Splitting Up is Complicated – Crash Course Biology #12

1) Develop a logical argument as to why cells in YOUR body need to undergo **mitosis** in order for YOU to survive.

____
____
____
____
____

2) Explain how **DNA** in the **nucleus** of a cell is organized in order to keep a human alive.

____
____
____
____

3) Explain what a cell does during **interphase** by clarifying why the cell needs **chromatin** to replicate itself. *Include the process of* ***centrosome duplication*** *inside of your answer.*

____
____
____
____
____

4) Analyze the process of **prophase** inside of **mitosis**. *Be sure to include a brief analysis of the functions of chromatids, centromeres, the nuclear envelope, and microtubules during* ***prophase****.*

____
____
____
____
____
____

5) Analyze the process of **metaphase** (after phase). *Be sure to include a brief analysis of the functions of* ***motor proteins****, and explain why the* ***chromosome****s line up in the center of the cell.*

____
____
____
____
____
____
____

6) Evaluate the importance of scientist Walther Fleming's (1843-1905) observation of a salamander **gill cell** in 1878.

____

____

____

____

____

____

____

7) Explain the contribution Professor Tomomi Kiyomitsu made to the understanding of **mitosis** during his observations of the motor protein **dynein** while he was a postdoctoral student at **MIT** in 2012.

____

____

____

____

____

____

____

____

____

____

____

8) Analyze the process of **anaphase** where **chromosomes** split from their copies.

____

____

____

____

____

____

____

____

____

____

9) Analyze the process of **telophase** when **genetic** material is rebuilt in a cell.

____

____

____

____

____

____

____

____

Name____________________
Period__________________
Date____________________

Meiosis: Where the Sex Starts – Crash Course Biology #13

1) Clarify the reasons why YOU cannot clone YOURSELF by just undergoing **mitosis** even though the vast majority of YOUR cells can clone themselves.

____
____
____
____
____

2) Explain the function of human **haploid cells** during the process of **meiosis**.

____
____
____
____
____

3) Clarify the major similarities and differences between **mitosis** and **meiosis**.

____
____
____
____
____

4) Differentiate between the different types of **gametes** created by **diploid cells** known as **primary oocytes** (aka **primary spermatocytes**).

____
____
____
____

5) Explain what happens in **meiosis** during **Prophase I**.

____
____
____
____
____

6) Clarify why **homologous recombination** is the whole point of **meiosis.** *Be sure to point out how that is different than **mitosis**.*

____
____
____
____
____

7) Connect how the process of **recombination** combines both the father and mother's **DNA** in order to create a new human being.

8) Apply concepts to understand why **meiosis** is good for **natural selection**.

9) Investigate and report on the 23rd chromosomes (aka **sex chromosomes**) which have difficulty going through the process of recombination in **meiosis**.

10) Explain how two **haploid cells** are created during round 1 of **meiosis**.

11) Clarify how two **diploid cells** change to four **haploid cells** during the second stage of **meiosis**.

12) Clarify the egg-making process during **meiosis**. *Be sure to include the function of **polar bodies** in both humans and plants inside of your answer.*

Name____________________
Period__________________
Date____________________

Natural Selection – Crash Course Biology #14

1) Explain how **natural selection** allows individuals to thrive and multiply as well as change genetically and evolve over time.

____
____
____
____

2) Explain the reasons why Charles Darwin (1809-1882), author of <u>On the Origin of the Species by Natural Selection</u>, came to understand the process of **natural selection**.

____
____
____
____
____

3) Evaluate the survival of the finch species by **adaptation** and **relative fitness** as observed by Darwin on the Galapagos Island which would become central to his idea of **natural selection**.

____
____
____
____
____
____
____
____
____

4) Differentiate between the four basic principles of **natural selection** based on Darwin's **observations**:

   a) Variations of Phenotypes

   ____
   ____
   ____
   ____

   b) Heritable

   ____
   ____
   ____
   ____

c) "The struggle for existence," -Darwin

d) Survival & Reproduction Rates

5) Apply concepts and explain why **natural selection** must take place.

6) Investigate and report on why certain **populations** of moths in London, England during the 19th century survived longer, such as that observed in the ratio of surviving black moths to peppered moths. *Be sure to include why this is a good example of **natural selection*** in your explanation.

7) Connect the reasons as to why Darwin would be concerned about the **genetic makeup** of his own children after observing the effects of **crossbreeding** and **inbreeding** in plants and animals inside of his scientific studies.

8) Differentiate why it is important not only to understand an organism's changing **phenotype** (physical form) but also its **genotype** (genetic form).

9) Explain the differences between **alleles** and **genes**.

10) Compare the different modes of selection that occur during **natural selection**:
    a) Briefly analyze the reasons why **directional selection** has caused the necks of giraffes to become long over time.

    b) Briefly analyze the reasons why **stabilizing selection** has caused the survival rate for human babies to be higher in the middle weight range rather than in the light or heavy weight range.

    c) Clarify the reasons why **disruptive selection** may take place in lakes by analyzing the epidemic of a yeast parasite inside of tiny crustaceans called Daphnia in 2008.

11) Outline how pressures on **populations** can come from **sexual selection** and **artificial selection** and not just from environmental pressures.
    a) **Sexual selection**

    b) **Artificial selection**

Name____________________
Period___________________
Date____________________

Speciation: Of Ligers & Men – Crash Course Biology #15

1) Briefly evaluate the success of **Homo Sapiens** who (over time) beat out **hominid** rivals **Homo Erectus** and **Homo Habilis** to populate the world.

____
____
____
____

2) Briefly explain why YOU think a **species** must be able to produce **fertile offspring**.

____
____
____
____
____

3) Develop a logical argument as to how a liger (lion + tiger) animal **hybrid** is created when all ligers are sterile.

____
____
____
____

4) Clarify how **evolution** has led to the creation of new **species**.

____
____
____
____
____

5) Analyze how **reproductive isolation** has led to **speciation** inside of a population of animals who used to share the same **genetic traits**.

____
____
____
____
____

6) Clarify how a mule is created, and then explain why a mule can't mate with another mule to create more mules due to its **post-zygotic** isolation.

____
____
____
____
____

7) Explain how a change in an organism and its chances of mating to produce fertile offspring could be influenced by **pre-zygotic** isolation shifts such as changes in **behavioral** or **geographic** factors.

   A) **behavioral** changes

   ____
   ____
   ____
   ____
   ____

   B) **geographic** changes

   ____
   ____
   ____
   ____
   ____

8) Hypothesize why a certain set of conditions could cause a species to diverge into two new species in a process called **allopatric speciation**.

   ____
   ____
   ____
   ____
   ____
   ____
   ____
   ____

9) Analyze an example of super-quick **sympatric speciation** that occurred inside of the finch population on the Galapagos islands during the 20th century.

   ____
   ____
   ____
   ____
   ____
   ____
   ____
   ____

10) Differentiate between **allopatric speciation** and **sympatric speciation**.

   ____
   ____
   ____
   ____
   ____
   ____
   ____

Name___________________
Period__________________
Date___________________

Animal Development: We're Just Tubes – Crash Course Biology #16

1) Evaluate the effectiveness of classifying animals into **animal phyla** in order to differentiate between simple organisms like sea sponges to very complex organisms such as a giraffe.

____
____
____
____
____

2) Formulate a hypothesis as to why a sea sponge, which has no organs, can **reconstitute** itself even after being put into a blender.

____
____
____
____
____
____

3) Explain how an animal's complexity is determined by the layers of **tissue** an organism creates in its first couple of hours as a **zygote**.

____
____
____
____
____
____

4) Analyze the formation of a **morula** and then a **blastula** from a **zygote** soon after the union of the sperm and the egg.

____
____
____
____
____
____

5) Differentiate between animals that have **radial symmetry** and animals that have **bilateral symmetry**.

____
____
____
____

6) Explain what occurs during **gastrulation,** or the forming of the **digestive tract,** which takes place in cells after **cleavage**.

7) Differentiate between a **protostome** and **deuterostome**, and then explain how they develop in the early **embryo** stages.

8) Analyze the differences between less complex **diploblastic** organisms that have only two germ layers and the more complex **triploblastic** organisms.

9) Differentiate between the functions of the **endoderm** (digestive tract), the **ectoderm** (skin), and the **mesoderm** (muscles, circulatory system, reproductive system, bones).

10) Explain why Ernst Haeckel (1834-1919) was not well liked by other scientists of his era by clarifying how his **recapitulation theory** was debunked.

Name___________________
Period__________________
Date___________________

Evolutionary Development: Chicken Teeth – Crash Course Biology #17

1) Briefly explain the science of **evolutionary biology**.

____
____
____
____

2) Briefly explain how **gap genes** influence the **blastula** into organizing an animal's development.

____
____
____
____
____
____

3) Compare the functions of **homeobox genes** and **regulatory genes** to head architects in charge of the construction of a building after the **embryo** is developed.

____
____
____
____
____
____
____
____

4) Hypothesize the reasons why **regulatory genes** and **gene products** wait inside of the unfertilized egg to tell **embryonic cells** what to do.

____
____
____
____
____
____

5) Clarify the function of **Hox genes** by pointing out the similarities and the differences along the **head-tail axis** of both a mouse and human **embryo**.

____
____
____
____
____

6) Explain what occurs during **gastrulation,** or the forming of the **digestive tract,** which takes place in cells after **cleavage**.

____

____

____

____

____

____

____

7) Differentiate between a **protostome** and **deuterostome**, and then explain how they develop in the early **embryo** stages.

____

____

____

____

____

____

8) Analyze the differences between less complex **diploblastic** organisms that have only two germ layers and the more complex **triploblastic** organisms.

____

____

____

____

____

____

9) Differentiate between the functions of the **endoderm** (digestive tract), the **ectoderm** (skin), and the **mesoderm** (muscles, circulatory system, reproductive system, bones).

____

____

____

____

____

____

____

____

10) Explain why Ernst Haeckel (1834-1919) was not well liked by other scientists of his era by clarifying how his **recapitulation theory** was debunked.

____

____

____

____

____

____

Name___________________
Period_________________
Date___________________

Evolutionary Development: Chicken Teeth – Crash Course Biology #17

1) Briefly explain the science of **evolutionary biology**.

2) Briefly explain how **gap genes** influence the **blastula** into organizing an animal's development.

3) Compare the functions of **homeobox genes** and **regulatory genes** to head architects in charge of the construction of a building after the **embryo** is developed.

4) Hypothesize the reasons why **regulatory genes** and **gene products** wait inside of the unfertilized egg to tell **embryonic cells** what to do.

5) Clarify the function of **Hox genes** by pointing out the similarities and the differences along the **head-tail axis** of both a mouse and human **embryo**.

6) Do YOU think it is **ethical** for a scientist to splice **genes** and reconstitute them in order to make a **chimera?** (mixture of two animals, insects, etc.)? *Why or why not? Defend your answer.*

7) Hypothesize as to why **evolution** and **genetic mutations** sometimes happen really fast.

8) Cite evidence inside of the **fossil record** as to why birds don't have teeth even though many of their ancestors such as the therapod dinosaurs had teeth. *Include the role* ***regulatory gene mutation*** *plays in this* ***progressive adaptation****.*

9) Apply concepts to explain why some **genes** can cause chickens to be born with teeth, some snakes can be born with legs, and some cave fish can be born with eyes.

Name___________________
Period__________________
Date___________________

Population Genetics: When Darwin Met Mendel – Crash Course Biology #18

1) Briefly explain how the field of **population genetics** (**Pop-Gen**) demonstrates how **genetics** and **evolution** influence each other.

____
____
____

2) Explain how a **population** of birds can grow to become a larger **population** of birds.

____
____
____

3) Connect the factors that cause changes in **allele frequency** which in turn drive **evolution**.

   A) **Natural Selection** (**alleles** for fitter organisms)

   ____
   ____
   ____
   ____
   ____

   B) **Sexual Selection** (nonrandom mating, **alleles** for more sexually attractive organisms)

   ____
   ____
   ____
   ____
   ____
   ____

   C) **Mutation** (new **alleles** pop up from mistakes in **DNA**)

   ____
   ____
   ____
   ____
   ____

   D) **Genetic Drift** (changes in **allele frequency** due to random chance)

   ____
   ____
   ____
   ____
   ____
   ____

E) **Gene Flow** (changes in **allele frequency** due to mixing with new genetically different **populations**)

____
____
____
____
____

4) Analyze the **Hardy-Weinberg Principle**, developed by Godfrey Hardy (1877-1947) and Wilhelm Weinberg (1862-1937), which explains how **Mendelian genetics** works at the scale of a whole **population**.

____
____
____
____
____
____
____

5) Briefly explain the crux of the **Hardy-Weinberg equilibrium**.

____
____
____
____

6) Clarify how the **Hardy-Weinberg equation** has strict requirements in order to happen.

a) NO **natural selection**

____
____
____

b) NO **sexual selection**

____
____
____

c) NO **mutations**

____
____
____

d) Gigantic **population** size

____
____
____

e) NO **gene** flow

____
____
____

7) a) Calculate the relationship between the expressed **phenotype** and **frequency** of genes (**dominant alleles** and **recessive alleles**) in the ratio of wet to dry earwax of a **population** of 100 humans on a deserted island if q^2 is 0.09. *(q^2 represents that 90% of the* ***population*** *on the island has wet ear wax)*

____/100 =____% Ww

p+q= ______. $(p+q)^2$ = _______

Hardy-Weinberg Equation $P^2 + 2pq + q^2 = 1$

$q^2 = \sqrt{.09}$ = _________

_______+_______ =1

√_______=_______

Once the math is figured out, go back and plug the math into your equations in order to find out how many ***heterozygotes*** *there are living on the island.*

___X___X____=_______

b) Explain which members of the **population** displays a **homozygous dominant genotype** and a **heterozygous genotype** in YOUR answer.

____

____

____

____

____

____

____

____

____

____

____

____

____

____

____

____

____

____

Name____________________
Period__________________
Date____________________

Taxonomy: Life's Filing System – Crash Course Biology #19

1) Explain how biologists use the **taxonomic system** (aka the **phylogenetic tree**) to classify all the organisms on Earth. *(kingdom, phylum, class, order, family, genus, species)*

____
____
____
____

2) Critique whether or not YOU think that Carl Linnaeus (1707-1778) author of Systema Naturae (1735) was correct in his adoption and use of **morphology** to classify plants and animals.

____
____
____
____
____
____

3) Why do YOU think we still use Linnaeus's **morphology**-based system of **classification** from the 18th century even though most of us don't speak Latin?

____
____
____
____
____
____

4) Analyze the reasons why a new **taxon** called **domain** was added to the start of the kingdom, phylum, class, order, family, genus, species tree.

____
____
____
____
____
____

5) Explain the difference between **taxon** and **species** in relation to the **classification** of organisms.

____
____
____
____

6) Differentiate between three different types of domains: **Bacteria**, **Archaea**, and **Eukarya**.
   a) **Bacteria**

   b) **Archaea**

   c) **Eukarya**

7) Clarify the differences between the kingdoms of **Protista**, **Fungi,** **Plantae**, and **Animalia**.
   a) **Protista**

   b) **Fungi**

   c) **Plantae**

   d) **Animalia**

8) Briefly analyze an organism of CHOICE and follow it all the way through the **taxa** from kingdom to species.

Name____________________
Period___________________
Date_____________________

Evolution: It's a Thing – Crash Course Biology #20

1) Explain how **evolution** works and clarify why it's not up for debate as to whether or not **evolution** actually occurred.

2) Briefly explain how the **theory of evolution** incorporates a large number of phenomena all at once.

3) Connect how **fossils** are both similar yet different to the organisms which are alive today.

4) Cite evidence from the **fossil record** which proves that the ancestors of whales used to walk on their hind legs.

5) Analyze how **homologous structures** found in shared **characteristics** between organisms such as those between a dog, a bat, and a whale shows how they are all related.

6) Clarify how, if life is ever to be found on Mars, the existence of **RNA** (**ribonucleic acid**) inside of its cells would be a good test if it were actually extraterrestrial or not.

____
____
____
____
____
____

7) Explain how an animal's **biogeography** is explained by the **Theory of Evolution**.

____
____
____
____
____
____
____

8) Hypothesize as to why the highest **concentration** of marsupials (kangaroos, koalas, and wombats) are found in Australia and not in North America.

____
____
____
____
____
____

9) Cite **biogeographical** evidence as to how Darwin's finches evolved and survived on different Galapagos Islands as separate **species** on each island due to **natural selection**.

____
____
____
____
____
____

10) Analyze how **direct observation** can prove the **Theory of Evolution** based on the **microevolution** of a number of organisms studied in the 20th and 21st centuries.

____
____
____
____
____
____
____
____
____

Name___________________
Period__________________
Date___________________

Comparative Anatomy: What Makes Us Animals – Crash Course Biology #21

1) Briefly explain why scientists study **comparative anatomy**.

____

____

2) Pinpoint how **locomotion** can be used to differentiate between **Kingdom Animalia** and **Kingdom Plantae**.

____

____

____

____

3) Explain how **heterotrophs** get energy from eating other life forms.

____

____

____

____

4) Draw conclusions as to how **convergent evolution** could cause the bodies of highly different animals such as a tuna, a penguin, and a seal to all have fins and similar-looking sleek fusiform bodies that can move through water.

____

____

____

____

____

____

5) Why do YOU think **convergent evolution** occurs in birds, fish, and mammals?

____

____

____

____

____

____

6) Explain how Thomas Henry Huxley (1825-1895) connected **paleontology** and **biology** together by hypothesizing the link between dinosaurs and birds.

____

____

____

____

____

____

7) Explain why animals need to **digest** food in order to create energy for themselves.

8) Analyze the differences between the four types of **primary tissues** in the human body:
   a) **Epithelial tissue** (layer covering every organ)

   b) **Connective tissue** (collagen protein)

   c) **Muscle tissue** (actin, myosin)

   d) **Nerve tissue** (neurons, glial cells)

9) Explain how sponges are different than other animals.

Name__________________
Period_________________
Date___________________

Simple Animals: Sponges, Jellies, & Octopuses – Crash Course Biology #22

1) Explain why **tissue complexity** in the kingdom animalia determines the complexity of an animal. *Include how this is determined during the* ***embryonic phase****.*

____
____
____
____
____
____

2) Briefly report on the history of **sponges** and then explain why they are still classified as animals.

____
____
____
____

3) Analyze the reasons why a **diploblast** is the oldest living ancestor of all true animals.

____
____
____
____
____

4) Evaluate the reasons why the phylum Platyhelminthes (soft **unsegmented worms** which include flatworms, planaria, tapeworms, and flukes) are such an important revolutionary breakthrough in the evolution of Kingdom Animalia.

____
____
____
____
____
____
____
____
____

5) Clarify the function of the **coelom** in aiding the development and function of **internal organs** inside of an organism.

____
____
____
____

6) Evaluate the importance of the **Cambrian Explosion** (541-485 million years ago) in the history of life on earth.

____

____

____

____

____

____

7) Explain the reasons why the increase of **oxygen levels** and the change of ocean chemistry during the **Cambrian Era** (541-485 million years ago) aided in the **population explosion** on earth during this time.

____

____

____

____

____

____

____

8) Differentiate between **pseudocoelomates** (unsegmented roundworms) and true **coelomates** (arthropods, molluscs, chordates).

____

____

____

____

9) Briefly explain how parasitic **hookworms** feed off of humans.

____

____

____

10) Clarify how animals from phylum **Rotifera** are more complex than **nematodes**.

____

____

____

____

11) Point out the similarities and differences between the four different types of **molluscs**: chitons, snails, bivalves, and octopi/squid.

A)____

____

B)____

____

C)____

____

D)____

____

Name___________________
Period_________________
Date___________________

Complex Animals: Annelids & Arthropods – Crash Course Biology #23

1) Analyze how **segmentation** and anatomically identical units in phyla Annelida and Arthropoda indicates an animal's complexity.

____
____
____
____
____
____
____

2) Cite evidence as to why **segmentation** has proven to be incredibly useful in humans from an evolutionary perspective.

____
____
____
____
____
____
____

3) Compare the **synapomorphy** found in earthworms in phylum Annelida to the bodily construction found in flatworms and nematodes in the phylum Nematoda.

____
____
____
____
____
____
____

4) Draw conclusions between the **plesiomorphies** of worms in the phylum Annelida in order to make connections as to how they all have a common ancestor.

____
____
____
____
____
____
____
____

5) Differentiate between the three different classes of **synapomorphy** found inside of the phylum Annelida.
   a) Oligochaetes (earthworms)

      ____
      ____
      ____
      ____
      ____

   b) Hirudinea (leeches)

      ____
      ____
      ____
      ____
      ____

   c) Polychaetes (bristly worms- marine species)

      ____
      ____
      ____
      ____
      ____

6) Analyze the **synapomorphies** that make members of the phylum Arthropoda (insects, centipedes, scorpions, spiders, lobsters, and crabs) similar to each other.
   a) **Segmented bodies**

      ____
      ____
      ____
      ____
      ____
      ____

   b) **Exoskeleton**

      ____
      ____
      ____
      ____
      ____
      ____

   c) **Paired** and **jointed appendages**

      ____
      ____
      ____
      ____
      ____
      ____

Name___________________
Period__________________
Date___________________

Chordates – Crash Course Biology #24

1) Analyze the reasons why the presence of a **notochord** makes **chordates** different from other animal phyla.

____
____
____
____
____

2) Differentiate as to how the presence of a **dorsal hollow nerve cord** makes **chordates** different from other animal phyla that have a **ventral solid nerve cord**.

____
____
____
____
____

3) Explain what the **pharyngeal gill slits**, or pouches in human **chordates**, develop into after humans are born.

____
____
____
____

4) Briefly analyze the function of the **post-anal tail** in aquatic **chordate** animals.

____
____

5) Explain the reasons why sea squirts in the subphylum **Urochordata** are part of the phylum **Chordata**.

____
____
____
____
____

6) Explain why YOU think a hard backbone in the **chordate** subphylum **Vertebrata** has allowed for an explosion in its diversity.

____
____
____
____
____

7) Analyze the reasons why YOU think that teeth began to emerge inside of **vertebrates**.

8) Explain why there has been so much hype after a live **Coelacanth** fish was discovered off the coast of South Africa in 1938.

9) Evaluate how **tetrapods** such as frogs adapted for survival by not only being the first creatures to develop a **three-chambered heart** but also the first to lay **amniotic eggs**.

10) Clarify how reptiles such as lizards, snakes, and turtles do not actually have cold blood but are **ectothermic**.

11) Evaluate the advantages birds have over other animals as they are able to regulate their **body temperature** since they are **endotherms** and have a **four-chambered heart**.

12) Analyze the advantages that the class **Mammalia**, which includes dogs and humans, has as they are not only **endothermic** but also are able to avoid dangers by developing first in their mother's body.

Name___________________
Period_________________
Date___________________

Animal Behavior – Crash Course Biology #25

1) Compare the differences between an **internal stimulus** and an **external stimulus** by analyzing the feeding of a cat as an example.

____
____
____
____
____

2) Evaluate how an animal's **behavior** could contribute to its **survival**. *Pick an animal of choice (ex. YOUR dog, cat, fish, lizard) and evaluate how its* ***behavior*** *contributes to its* ***survival****.*

____
____
____
____
____
____

3) Differentiate between the animal functions of **morphology** and **physiology**.

____
____
____
____
____

4) Analyze how the **evolutionary advantage** of having a long neck, such as a giraffe, could contribute to its overall **survival** in the passing of its **genes** to its **offspring**.

____
____
____
____
____

5) Briefly explain how **adaptive behavior** -in an animal (of choice)- could contribute to its survival by analyzing the **stimulus** that causes the **functions** that this behavior serves.

____
____
____
____
____
____

6) Examine and report on the **proximate causes** that a male Siberian hamster exhibits in stimulating it to mate with a female Siberian hamster. *Include the* ***function*** *that this* ***behavior*** *serves.*

7) Evaluate the reasons why **pheromones** are the **ultimate causes** behind the **behavior** of the male Siberian hamster in choosing to find a **mate** even in adverse conditions.

8) Explain how **imprinting** can cause an animal to develop **social bonds** with its master even if the master is human.

9) Analyze the two types of **behavior** which shows how **natural selection** acts on behavior in the world:
   a) **Foraging** (finding and eating food) -**Optimal foraging theory** (OFT)

   b) **Sexual selection** (having sex with the right **mate**)

Name_______________
Period_______________
Date_______________

The Nervous System – Crash Course Biology #26

1) Analyze the two main departments that function as the bureaucracy of **neurons** inside of YOUR **nervous system**:
   a) **Central nervous system** (brain, spinal cord)

      ____
      ____
      ____
      ____

   b) **Peripheral nervous system** (all nerves outside brain and spinal cord)

      ____
      ____
      ____
      ____

2) Compare how the **central nervous system** and the **peripheral nervous system** interact with each other.

   ____
   ____
   ____
   ____

3) Differentiate between the functions of **afferent** and **efferent** neurons in the **central nervous system**.

   ____
   ____
   ____
   ____
   ____

4) Explain why **efferent neurons** are called **motor neurons** in the **peripheral nervous system.**

   ____
   ____
   ____
   ____

5) Differentiate between the functions of the two different **efferent** (motor) **neurons**: the **somatic nervous system** and the **autonomic nervous system**.
   a) **Somatic nervous system** (all the things you think about doing)

      ____
      ____
      ____
      ____

b) **Autonomic nervous system** (heartbeat, digestion, breathing, saliva production, organ functions)

6) Clarify how the **reflex loop** controls what is going on inside your **somatic nervous system** when YOU touch a hot stove rather than a response controlled by YOUR brain.

7) Compare the responsibilities of the **sympathetic division** to the responsibilities of the **parasympathetic division** in the **automatic nervous system**.

a) **Sympathetic division** (fight or flight)

b) **Parasympathetic division** (chilling out)

8) Analyze the structure of a typical **neuron** and then explain the function of its **dendrites** and **axons**.

a) **Neuron**

b) **Dendrites**

c) **Axons** (insulated with **myelin**)

9) Clarify the function of the little bits of exposed **nerve** along the **axon** called the **Nodes of Ranvier**.

10) Explain how **saltatory conduction** allows for signals to travel along a **neuron**.

11) Briefly analyze how **dendrite cells** communicate with each other through **synapses**.

12) Compare how the transport process travels down the **concentration gradient** of a regular **cell membrane** -aka the **sodium-potassium pump** - and is somewhat similar to the process found in the **membrane potential** inside the cell of a **neuron**.

13) Briefly explain what happens during an **action potential** in a **nerve cell**. *Include when the internal charge of a **neuron** reaches a certain **threshold** to become active.*

14) Explain what happens when YOU eat a pizza by analyzing how **neurons** travel from YOUR taste buds up through YOUR **spinal cord** and into YOUR **brain** and back out again.

Name___________________
Period__________________
Date___________________

Circulatory & Respiratory Systems – Crash Course Biology #27

1) Briefly explain the reasons why the human body has a **respiratory system** and a **circulatory system**.

____
____
____
____
____

2) Cite evidence as to how some animals and amphibians can take in **oxygen** (O_2) without lungs using a process called **simple diffusion**.

____
____
____
____

3) Analyze how a **fish** uses its **gills** to breathe **oxygen** and excrete **carbon dioxide** (CO_2).

____
____
____
____

4) Apply concepts and explain the reasons as to why -when a **red tide** (**algae blooms** which deplete **oxygen** from the water) occurs - that so many fish die off in the ocean and wash up on the beaches.

____
____
____
____

5) Clarify how the **bronchioles** in human **lungs** covert **oxygen** from the atmosphere and make it available for use in a human's **circulatory system**.

____
____
____
____

6) Explain the role **alveoli** plays in exhaling the converted **carbon dioxide** out of a human's body.

____
____
____
____
____

7) Analyze how **pressure** in the **lungs** work like a pump so that a human can breathe in **oxygen** and exhale **carbon dioxide**.

8) Differentiate between the **circulatory system** and the **respiratory system**.

9) Clarify the role of the heart in the **circulatory system**.

10) Explain the role that **arteries** play in maintaining **pressure** and delivering **oxygen** inside of the human body.

11) Investigate the role that **veins** play in carrying **deoxygenated blood** back to the heart.

12) Explain the role that **pressure** inside of **veins** plays inside of the **circulatory system**.

13) Differentiate between the **inferior vena cava** and the **superior vena cava**.

Name___________________
Period__________________
Date___________________

The Digestive System: Crash Course Biology #28

1) Explain how the **digestive tract** of a fly is specifically adapted to its **feeding behavior**.

____
____
____

2) Apply concepts and explain why the **digestive tract** among **vertebrates** such as dogs are extremely short due to its feeding habits of eating rotten meat.

____
____
____
____

3) Explain why **vertebrates** such as cows need **microorganisms** in their **guts** in order to digest grass.

____
____
____
____
____
____
____

4) Clarify why chewing is key to the digestive systems of humans, and then explain how humans use **enzymes** and **acids** in order to aid in their digestion of their food.

____
____
____
____
____
____
____
____
____

5) Analyze the similarities and differences between the **digestive system** and **circulatory system** of a human.

____
____
____
____
____
____

6) Explain why it is necessary to chew bread in order to activate the **salivary amylase** enzyme in the human mouth.

7) Clarify the role **gastric juice** plays in breaking down food inside of the **stomach** of the human body.

8) Analyze why the role of the **small intestine** works so well in **absorbing nutrients** and **breaking down fats** as well as neutralizing **gastric juice**.

9) Clarify the role of **bile** in aiding the **small intestine** break down **fat molecules** during **digestion**.

10) Assess why YOU think that the job of the **large intestine** is to remove most of the water and **bile salts** from the **chyme** (pulpy acidic fluid of partly digested food) and **bile** travelling through your body.

11) Draw conclusions as to the function of the human **appendix** which is located at the end of the **cecum** (the first region the **large intestine**).

Name____________________
Period__________________
Date____________________

The Excretory System: From Your Heart to the Toilet – Crash Course Biology #29

1) Draw conclusions as to the reasons why **ammonia** (NH_3), which comes from breaking down proteins during the **digestion** process, is converted into **urea** or **uric acid** in the **liver**.

____

____

____

____

____

2) Explain why drinking **water** (H_2O) is necessary to dissolve **urea** (CH_4N_2O).

____

____

____

____

____

3) Analyze the white **uric acid** ($C_5H_4N_4O_3$) and the brown parts of bird poop in order to explain why the bird poop looks the way it does.

____

____

____

____

____

4) Evaluate how different parts of the **kidneys** (i.e. **Glomerulus**, **Bowman's capsule**, **proximal convoluted tube**, and the **Loop of Henle**) maintain the **blood pressure** as well as the levels of water and dissolved materials in the **excretory system** of the human body. *Include how **nephrons** in the **kidneys** convert fluid from **blood** to **urine**.*

   a) **Glomerulus -nephron** starting point

   ____

   ____

   ____

   ____

   ____

   b) **Bowman's capsule** -filtrate

   ____

   ____

   ____

   ____

   ____

c) **Proximal convoluted tubule** -osmo-regulation

d) **Loop of Henle**

d1)

d2)

d3)

5) Evaluate the effectiveness of the whole process of fluid conversion from the **kidneys** to excreted **urine** in the toilet.

6) Explain why a person urinates a lot after drinking **alcohol**. *Include the reasons why the **urine** is clear.*

Name__________________
Period__________________
Date__________________

The Skeletal System: It's ALIVE! – Crash Course Biology #30

1) Briefly explain the reason why a **skeletal system** is vital for an animal's survival. (i.e. *What would happen if YOU didn't have a skeleton?*)

2) Clarify how jellies and worms are able to change shape and move with their **hydrostatic skeletons**.

3) Evaluate the pros and cons of humans having an **endoskeleton** rather than an **exoskeleton**.

4) Cite evidence as to why the Roman doctor **Galen** (129-216) was wrong in his assumptions about the human body.

5) Analyze new **bone structure** by investigating the composition of **cartilage** and by explaining the function of specialized cells called **chondrocytes**.

6) Explain the process of **ossification** in which **blood vessels** bring **osteoblasts** into the **cartilage** of the bone to form **calcium phosphate**.

7) Differentiate between the two layers of bones in humans: the **cortical bone** (compact bone) and the **trabecular bone** (spongy bone).
   a) **Cortical bone** (compact bone)

   b) **Trabecular bone** (spongy bone)

8) Explain how the **femur bone** in the leg increases in size as one grows from a child into an adult. *Clarify the role of the **diaphysis**, **epiphysis**, and **epiphyseal plate** in your answer.*

9) Identify and explain the role of the **pituitary gland** inside of the brain. *Include its managing of **bone growth** by utilizing a process known as **bone remodeling**.*

10) Differentiate between the roles of **osteoblasts** and **osteoclasts** in breaking down and reforming bones in YOUR body.

Name____________________
Period__________________
Date____________________

Big Guns: The Muscular System – Crash Course Biology #31

1) Briefly explain the three different types of muscles in the human body: the **cardiac muscle**, the **smooth muscle**, the **skeletal muscle**.
   a) **Cardiac muscle** (heart muscle)

   ____
   ____
   ____

   b) **Smooth muscle** (involuntary processes)

   ____
   ____
   ____

   c) **Skeletal muscle** (ex. gluteus maximus, masseter, abductor pollicis brevis)

   ____
   ____
   ____
   ____

2) Analyze both the composition and function of **tendons** in aiding with the contraction and relaxing of **muscles** inside of the human body.

____
____
____
____
____

3) Compare the **skeletal muscle** in YOUR arm to a rope and then explain the structure of that muscle.

____
____
____
____
____

4) Investigate how the specialized jobs that the **muscle fibers** inside of the layers of **muscle fascicles** perform inside of a chicken breast in order to make **proteins**.

____
____
____
____
____
____
____

5) Differentiate between the function of **myofibrils** and **sarcomeres** inside of the muscle system.

6) Clarify how the protein **actin**, a type of **myofilament**, is responsible for the contracting and relaxing of YOUR muscles as well as being involved in the motion of other types of processes and **organelles** inside of cells.

7) Evaluate the reasons why it took so long for humans to understand that **muscle contraction** was caused by the movement of one **protein** over another.

8) Explain how the **sliding filament model** works when the **muscle cell** is at rest. Then clarify how muscle contraction starts by relating the proteins **tropomyosin** and **troponin** to chaperons at a middle school dance that protect the **actin** from contact with the **myosin**.

9) Briefly explain how **rigor mortis** works inside of **muscle cells** after somebody dies.

Name____________________
Period__________________
Date____________________

Your Immune System: Natural Born Killer – Crash Course Biology #32

1) Analyze the two ways in which the human body fights against **pathogens**: **Innate immunity** and **acquired immunity**.
   a) **Innate immunity**

   ____

   ____

   ____

   b) **Acquired immunity**

   ____

   ____

   ____

2) Explain the reasons why the **skin** and the **mucous membranes** are the first line of defense in defeating **pathogens** that enter into the body.
   A) **skin**

   ____

   ____

   ____

   ____

   ____

   B) **mucous membranes**

   ____

   ____

   ____

   ____

   ____

   ____

3) Briefly explain what **mucus** does in keeping illness away from the human body.

   ____

   ____

   ____

4) Evaluate the effectiveness (pros and cons) of the release of **histamine** when **mast cells** issue an **inflammatory response**.

   ____

   ____

   ____

   ____

   ____

   ____

5) Explain what happens during the allergic reaction created during an **inflammatory response**, and then show how **antihistamines** are used to counter that reaction.

6) Clarify the function of **leukocytes** (white blood cells) inside of the human body. *Be sure to then explain how **diapedesis** occurs inside of a cell.*

7) Briefly explain how the process of **phagocytosis** ingests other cells and particles.

8) Investigate and report as to how **pus** forms from dead **neutrophils** in the human body.

9) Compare **macrophages** (large **phagocytes**) to body guards inside of the human **organs**.

10) Explain how **NK cells** (**Natural killer cells**) play a major role in the host rejection of both **tumors** and **virus-infected cells**.

____

____

____

____

____

11) Clarify how **dendritic cells** first eat up **pathogens** and then report back to the **acquired immune system.**

____

____

____

____

____

12) Draw conclusions as to how the **acquired immune system** first learns about **pathogens** and then creates an **antibody generator** to target its **antigens** to remove the **pathogen**.

____

____

____

____

____

____

13) Explain how **antibodies** contribute to the destruction of the **pathogens** that enter the human body.

____

____

____

____

____

14) Describe the roles of the two types of **lymphocytes** in the **acquired immune system** of the human body: **T Cells** and **B Cells**.

a) **Cell Mediated Response**- *Differentiate between the different types of **T Cells**.*

A1) **Helper T Cells** *-call shots (**antigen** presentation)*

____

____

____

____

____

____

____

____

____

A2) **Effector T Cells** -*steers immune response*

A3) **Memory T Cells** -*long term*

A4) **Cytotoxic T Cells** -*destruction of intracellular* ***pathogens*** *by* ***microphages***

b) **Humoral Response –B Cells** (type of white blood cell that secrete **antibodies**)

15) Describe how the **plasma/effector cells** create **antibodies** to destroy a **pathogen**.

Name____________________
Period__________________
Date____________________

Great Glands- Your Endocrine System: Crash Course Biology #33

1) Briefly explain the function of the **endocrine system** and how it influences **hormones** in the human body.

____
____
____
____
____
____

2) Clarify the differences between the **endocrine system** and the **nervous system** inside of the human body.

____
____
____
____

3) Differentiate between **endocrine glands** and **exocrine glands**.

____
____
____
____

4) Draw conclusions as to why all **glands** in the human body (i.e. **pituitary gland**, **thyroid**, **adrenal glands**, **pancreas**, **gonads**) have **blood vessels** connecting them directly to the **bloodstream**.

____
____
____
____

5) Briefly explain why **testosterone**, a **paracrine** regulator, only releases **hormone** molecules in the **testes**.

____
____
____
____

6) Clarify the three types of **signal receptors** inside of the human body: **steroids**, **peptides**, and **monoamines**.
   a) **Steroids**

   ____
   ____

b) **Peptides**

c) **Monoamines**

7) Compare the reaction of **lipids** to **signal receptors** such as **peptides** and **monoamines** to its reaction to **signal receptors** such as **steroids**.

8) Analyze how the **pituitary gland** inside a woman could cause that woman to start **lactating** if there is a nearby baby crying.

9) Clarify the function of **hormones** secreted by the **posterior pituitary gland** and made in the **hypothalamus**.
   a) **Oxytocin** (orgasms, pair bonding, anxiety, social recognition)

   b) **Antidiuretic hormone** (kidneys)

10) Analyze the function of the **thyroid** in the **endocrine system** of the human body. *Include the function of its communication system called the **negative feedback loop**.*

11) Design a scenario of choice in order to explain how the **hypothalamus** tells the **pituitary gland** to release the **adrenocorticotropic hormone** (ACTH) into the human body so that the **adrenal glands** are stimulated to make **epinephrine** (**adrenaline**).

____

____

____

____

____

____

____

12) Analyze how **epinephrine** works in the bloodstream to give that extra burst of energy to a human.

____

____

____

____

13) Design a scenario of choice to explain a time in which YOU remember when YOUR heart raced like crazy after YOUR body released **adrenaline** into YOUR **bloodstream**.

____

____

____

____

____

____

____

____

____

____

14) Explain the function of the **pancreas** in the **endocrine system** as well as how it works in regulating **glucose** levels inside of the human body. **Key terms to explain in the chemical conversion process= glucose=>cellular respiration=>insulin, glycogen=>adipose cells=>fat, energy=glucagon=>glycogen, fat=>glucose*

____

____

____

____

____

____

____

____

____

____

____

____

15) Connect how the two different types of **gonads** (sex glands), aka the **testes** and **ovaries**, make **sex hormones** as dictated by the **pituitary gland** in both males and females respectively.

a) **Testes** (male) – **androgens, testosterone**

b) **Ovaries** (female) – **estrogen, progestins**

16) Explain the reasons why **hormones** play a crucial role in determining the sex of a human.

17) Connect and explain why teenage boys experience a deepening of voice and an increase of hair production along with increased muscle and bone mass during the onset of **puberty**.

18) Connect and explain why **menstruation** and breast growth of teenage girls occurs during **puberty**.

Name___________________
Period_________________
Date___________________

The Reproductive System: How Gonads Go – Crash Course Biology #34

1) Briefly explain why some female spiders not only **mate** with a bunch of different male spiders but also store their **sperm** in different storage units.

____

____

____

2) Analyze the reasons why an animal needs to find another animal with a different type of **gamete** (sex cells) in order to have sex and reproduce.

____

____

____

3) Differentiate between the two different types of male and female **gametes**: **sperm** and **eggs**.

____

____

____

____

4) Clarify why garden snails are a **hermaphroditic** species.

____

____

____

____

5) Compare **reproduction** in the **plant kingdom** to **reproduction** in the **animal kingdom**.

____

____

____

____

____

6) Connect the reasons why female animals tend to be pickier about their **mates** while males are generally louder, larger, brighter, and more combative than females.

____

____

____

____

____

____

____

____

____

7) Differentiate between what determines whether or not an organism will have the **XX female chromosome** or the **XY male chromosome**.

____
____
____
____
____

8) Determine and then explain at what point in life a girl produces her **eggs** in contrast to when a boy produces his **sperm**.

____
____
____
____
____

9) Analyze the **secondary sex characteristics** of boys and girls which develop around puberty and then contrast these **secondary sex characteristics** against those found in other animals.

____
____
____
____

10) Analyze the function of female **reproductive structures** during the **menstrual cycle** of a human female. *Include the importance of the* ***endometrium*** *in your answer.*

____
____
____
____
____
____
____
____
____

11) Analyze the function of male **reproductive structures** in a human male. *Include the importance of the* ***testes, seminal vesicles, vas deferens, and semen*** *in your answer.*

____
____
____
____
____
____
____
____
____

Name____________________
Period___________________
Date_____________________

Old & Odd: Archaea, Bacteria & Protists – Crash Course Biology #35

1) Differentiate between **prokaryotic** (**unicellular**) and **eukaryotic** (**multicellular**) organisms by explaining a few of the characteristics of each.

____
____
____
____
____
____

2) Compare and then explain the similarities between **prokaryotic** and **eukaryotic** organisms. *Include the key terms **cytoplasm** and **ribosomes** in YOUR answer.*

____
____
____

3) Analyze the conditions of the early earth when **Archaean** life emerged during the **Archaean Eon** (4 billion years ago- 2.5 billion years ago) about 3.5-4 billion years ago.

____
____
____
____
____

4) Explain why YOU think the first **anaerobic** lifeforms that emerged during the **Archaean Eon** didn't need **oxygen** (O_2) to survive.

____
____
____
____
____
____

5) Analyze the reasons why YOU think some **terrestrial** scientists think that **methanogens** could be producing the small amounts of **methane** (CH_4) detected on the planet **Mars** even though its atmosphere contains only trace amounts of **oxygen** (O_2) gas (0.174%).

____
____
____
____
____
____
____

6) Clarify how **extremophiles** such as the **thermophiles** are different than other organisms in their ability to survive above the boiling point of water. **Boiling Point H_2O=212 Fahrenheit (F) or 100 Celsius (C).*

____

____

____

____

____

____

____

7) Investigate and explain how **bacteria** uses **horizontal gene transfer** to increase its **antibiotic** resistance by making it difficult at times for doctors to help some of their patients get better from an illness.

____

____

____

____

____

____

8) Assess how bacteria such as **Gram-positive bacteria** and **Gram-negative bacteria** can be both beneficial and detrimental to our lives. *Provide examples of each in YOUR answer.*

____

____

____

____

____

____

____

9) Briefly explain the importance of **cyanobacteria** inside of the **ocean food web** of fresh water and marine ecosystems.

____

____

____

____

10) Clarify how **protists**, or **protozoa**, such as **algae** and **amoebas** are more sophisticated than **prokaryotic** organisms such as **bacteria**.

____

____

____

____

____

____

____

Name____________________
Period___________________
Date____________________

The Sex Lives of Nonvascular Plants: Alternation of Generations – Crash Course Biology #36

1) Differentiate between the way water (**H_2O**) is transported in the **tissues** of **vascular plants** and then explain how **H_2O** is transported through **osmosis** in **non-vascular plants**.

____
____
____
____
____
____
____
____

2) Apply concepts to explain why **vascular plants** such as redwood trees are able to grow to an average height of 220 feet (67 meters).

____
____
____
____
____
____
____

3) Briefly analyze a few of the reasons why **non-vascular plants** such as moss have a **limited growth potential**.

____
____
____
____

4) Differentiate between the three types of **non-vascular plants**, aka **bryophytes**, today.
   a) **Liverworts**

   ____
   ____
   ____

   b) **Hornworts**

   ____
   ____
   ____

   c) **Mosses**

   ____
   ____
   ____

5) Describe how a **gametophyte** reproduces sexually in the **reproductive cycle** to produce **bryophyte sperm** and **eggs** which in turn give rise to an **asexual** second generation called the **sporophyte generation**.

6) Clarify how **spores** rely on their parent **sporophyte** for **water** (H_2O) and **nutrients** prior to their dispersal and subsequent continuance inside the circle of life.

7) Differentiate between how male and female **plants** reproduce their **sperm** and **eggs** during their **haploid** phase as **gametophytes**.
   a) **Male gametophyte – antheridia**

   b) **Female gametophyte – archegonia**

8) Analyze how adding **water** (H_2O), along with the process of **mitosis**, allows the **reproductive cycle** of **bryophytes** such as moss to work effectively to create more moss.

Name____________________
Period__________________
Date____________________

Vascular Plants = Winning! Crash Course Biology #37

1) Evaluate the reasons why **vascular plants** have had such extraordinary evolutionary success on planet Earth the last 700 million years.

____
____
____
____
____
____

2) Analyze how **conductive tissues** have aided **vascular plants** with their evolutionary success on the planet Earth.

____
____
____
____
____

3) Differentiate between the three types of **tissues** that define **vascular plants**:
   a) **Dermal tissues**

   ____
   ____

   b) **Vascular tissues**

   ____
   ____

   c) **Ground tissues**

   ____
   ____
   ____

4) Draw conclusions as to why **herbaceous** plants such as herbs and flowers are limited to the stage of **primary growth**.

____
____
____
____

5) Analyze the three ways in which plants such as trees experience a **secondary growth** and are afforded the ability to grow taller and wider due to their **woody tissues**.
   a) **Root**

   ____
   ____
   ____

b) **Stem – meristems**

c) **Leaf**

6) Clarify how the different sections of plants exploit **water** (H_2O) for their overall benefit.

a) **Dermal tissues**

b) **Epidermis**

b1) **cuticle**

b2) **Trichomes**

b3) **Root hairs**

b4) **Parenchyma** (flesh cells)

7) Explain how **negative pressure** pushes water (H_2O) upwards defying gravity through **evapotranspiration** inside the **xylem** of a plant.

8) Research and report the role that **transpiration** in plants plays in the **water cycle** moving water from the ground and into the atmosphere and back down to the ground again.

9) Clarify the function of **collenchyma** in celery.

10) Assess the advantages large woody plants have by analyzing the function of their **sclerenchyma cells** (bark). *Include how **sclerenchyma cells** change their function depending on whether or not the **climate** is warm and wet or cold and dry.*

11) Explain the role of **mesophyll** in **photosynthesis** by comparing the middle layer of the leaf to the bacon found in a BLT (bacon, lettuce, and tomato) sandwich.

12) Clarify the reasons why the **stomata**, located in the outer layer of a leaf, regulates the size and shape of the leaf.

13) Explain the pressure flow model for sugar **translocation** in the **phloem** tissue of a **plant cell.** *Include how **sap** plays a role in transporting **sugar** ($C_{12}H_{22}O_{11}$) throughout a plant.*

Name____________________
Period___________________
Date____________________

The Plants & The Bees: Plant Reproduction – Crash Course Biology #38

1) Explain the function of the **spores** that are released into the air by **vascular plants**.

____
____
____
____

2) Clarify the reasons why **nonvascular plants** are **gametophyte dominant** and have only one set of **chromosomes**.

____
____
____
____

3) Differentiate as to how **vascular plants** are **sporophyte dominant** whereas **nonvascular plants** are **gametophyte dominant** by identifying and explaining examples of each.

____
____
____
____
____
____
____

4) Analyze how the **spores** of ferns reproduce and then **germinate** to make new ferns.

____
____
____
____
____

5) Explain the evolutionary change from **spores** to **seeds** by identifying the role **gymnosperms**, such as coniferous plants (pine trees), played in this evolutionary advancement. *Be sure to identify and explain the reasons why a pine* ***cone*** *is NOT enclosed in an* ***ovule*** *like other seed-bearing plants.*

____
____
____
____
____
____
____
____

6) Connect the reasons why **Lodgepole evergreen trees** have evolved to take advantage of forest fires. *Be sure to explain how these trees take advantage of forest fires.*

7) Explain how the **mutualism** of flower-bearing **Angiosperm** plants and insects -such as bees- resulted in the **co-evolvement** of both leading to the pollination of more flowers.

8) Analyze how the **females** of **perfect flowers** work starting from the bottom up.

9) Clarify how **pollination** works to pollinate a flower to create seeds, fruit, and ultimately more flowers.

10) Investigate and explain the reasons why a zucchini is a **fruit** (and not a **vegetable**). *Include where the **fruit** is located on a strawberry in YOUR answer.*

Name__________________
Period__________________
Date__________________

Fungi: Death Becomes Them – Crash Course Biology #39

1) Cite evidence as to how **saccharomyces fungi** (brewer's yeast) plays a role in creating some food and drinks that humans consume.

____
____
____

2) Draw conclusions as to the reasons why some **fungi** such as the fungus **Claviceps** purpurea can cause the disease **ergotism** (aka St. Anthony's fire) in people and can even cause them to go crazy if consumed.

____
____
____
____
____

3) Explain what happened to bats in North America in 2006 due to a fungal disease called **white-nose syndrome** (WNS).

____
____
____
____

4) Analyze how **fungi** play a vital role in the **global food web**. *Be sure to include the role they play in converting **organic matter** back into soil.*

____
____
____
____
____

5) Evaluate the importance of the contributions that Louis Pasteur (1822-1895) made by showing the world how the process of **fermentation** breaks down sugars into alcohol using the **anerobic respiration** of fungi microorganisms such as **yeast**.

____
____
____
____
____
____
____
____
____

6) Differentiate between **heterotrophs** such as **fungi** and **heterotrophs** such as **humans**.

7) Compare how **chitin** found inside the cell walls of **fungi** differ from the **cellulose** of cell walls found in **plants**.

8) Evaluate how the **hyphae**, which make up the **mycelium** in **fungi**, are structured to maximize the absorption of food.

9) Examine three of the ways in which **fungi** can be classified by organizing the way in which they interact with other organisms and then analyzing examples of each:
   a) **Decomposers** -ability to break down **lignin** => **glucose**
      Ability to break down **proteins** => **component amino acids**

   b) **Mutualists** – **haustoria** & **plant roots** => **phosphates** = **mycorrhizae**

c) **Predators & parasites**

c1) **Predatory fungi** – **hyphae** & soil fungus Arthrobotrys

c2) **Parasitic fungi** – Ophiocordyceps (zombie ant fungus)

10) Analyze how **fungi** reproduce both sexually and asexually through a process called **plasmogamy** to create **spores** which develop into more **fungi**. *Include how their reproduction is different from the concept of male and female in both plants and animals.*

11) Analyze how **fungi** such as **mold** reproduce to create more **mold** on an orange, tomato, or a piece of bread.

12) Briefly analyze how brewers of **beer yeast** encourage **yeast** to create more **yeast** and ultimately beer.

Name__________________
Period_________________
Date__________________

Ecology – Rules for Living on Earth: Crash Course Biology #40

1) Construct a brief argument that connects how **ecology** explains the place of all biological organisms living on the earth.

____

____

____

2) Differentiate between organisms in a **population** and organisms in a **community**.

____

____

____

____

____

3) Connect the reasons why disturbing or upsetting an **ecosystem** can have disastrous results on that **ecosystem**. *Provide at least one example.*

____

____

____

____

____

4) Briefly assess the reasons why an organism would adapt to a general set of conditions as found inside of a **biome**. (I.e. **Land biomes** = tundra, taiga, temperate and deciduous forest, temperate and tropical rainforest, temperate grassland, chaparral, desert, savanna, and alpine. I.e. **Freshwater aquatic biomes** = lakes, rivers, and wetlands. I.e. **Marine biomes** = coral reefs, estuaries, and oceans.)

____

____

____

____

____

____

____

5) Differentiate between **biotic** and **abiotic** factors in a **biome**. *Include the factors that determine what an* ***ecosystem*** *will look like.*

____

____

____

____

____

____

6) Create and analyze scenario which proves how **enzymes** within YOUR body are most effective within a certain set of **temperatures**. *E.g. Do YOU study better in hot, warm, or cold environments? Or please create and analyze a scenario of choice.*

____

____

____

____

____

____

____

____

____

6b) Why Do YOU think **temperature** dictates the things certain organisms can do?

____

____

____

____

____

____

7) Explain the reasons why the search for food and water define every **biome** on the planet. *E.g. Explain what would happen to an animal or plant whose* ***biome*** *was originally in a* ***desert*** *and then they were placed them into a* ***rainforest****.*

____

____

____

____

____

____

____

____

8) Analyze why a biome's **physiognomy**, such as how wet, dry, hot, or cold it is, in effect determines the way it looks.

a) **Tropical rainforests**

____

____

____

____

____

b) **Tundra**

____

____

____

____

c) **Desert**

d) **Temperate grasslands**

e) **Temperate deciduous forests**

f) **Taigas/Coniferous forests**

g) **Marine**

9) Connect what would happen to a biome's **physiognomy** if it didn't have a lot of trees and had a lot of **grassland** instead.

BIOLOGY EXTRA ESSAYS SECTION

The Human Factor

1) How do YOU think humans are affecting **biomes** on the planet earth? **Hint-Go through each of the seven major* ***biomes*** *and then explain how human activity has affected each one.*

Ocean Life Sustainability Project

2) Research current levels of oxygen inside of the earth's oceans. Then make a decade by decade prediction on the sustainability of life in the oceans in the following centuries if current trends continue. Include data on current level of oxygen loss, cite evidence on past century levels of oxygen loss, and predict steps that would need to be taken to preserve the biological life forms in the ocean today.

Florida's Recurring Red Tide Events

3) YOU are a marine scientist studying the reasons surrounding the **red tide** events that are happening with greater frequency off the Southwest Coast of Florida. Hypothesize whether or not increased rainfall to the region due to global warming may be causing greater runoff and longer than normal **algal blooms** which cause the **red tide**. *Include marine animals which may be affected and then explain why months-long* ***red tide*** *events may be occurring with higher frequency than the past. Predict the future of the region based upon the information and data available in the present.*

Is Mold Good or Bad?

4) Analyze the reasons why old rye bread could be hazardous to a person's health.

The Moldy Orange Experiment

5) Observe and report the growth of mold on an orange over the course of two weeks. Take a picture daily and write notes on how quickly the mold grows and takes over the orange.

Day 1:

____

____

____

____

Day 2:

____

____

____

____

Day 3:

____

____

____

____

Day 4:

____

____

____

____

Day 5:

____

____

____

____

Day 6:

____

____

____

____

Day 7:

____

____

____

____

Day 8:

____

____

Day 9:

Day 10:

Day 11:

Day 12:

Day 13:

Day 14:

Which Hamburger Attracts More Mold and Why?

6) Observe and report the rate at which mold grows on a piece of cooked hamburger meat from the market versus how mold grows over a McDonald's hamburger over the course of two to three weeks. Take a picture daily of each hamburger and write notes on your observations of each burger. Once YOUR experiment is over, hypothesize why YOU think that mold grows faster on one hamburger than the other? Once you are done with your notes, research what constitutes each hamburger (what each hamburger is made out of).

Day 1:

____

____

____

____

Day 2:

____

____

____

____

Day 3:

____

____

____

____

Day 4:

____

____

____

____

Day 5:

____

____

____

____

Day 6:

____

____

____

____

Day 7:

____

____

Day 8:

Day 9:

Day 10:

Day 11:

Day 12:

Day 13:

Day 14:

Yummy Cheese!

7) Analyze the ways in which mold helps to make cheese.

Medical Mold

8) Analyze the ways in which Penicillin aids in curing many infectious diseases. *Include how it is derived from the mold Penicillium.*

The Ethics of Constructing a Chimera

9) Formulate an opinion as to whether or not creating a **human chimera** (animal +human) is ethical. After researching the subject, determine your stance on the subject. Evaluate the pros and cons of science's ability to create a **chimera**. Include how far YOU think scientists should go in the mixing together of human and animal **genes**.

b) After you complete your project, debate the **ethics** around creating a **chimera** with a partner of choice.

Be sure to state in your thesis/hypothesis as to when a line would be crossed for YOU during the genetic crossing of a human and an animal.

Thesis:

____

____

____

____

____

Counterargument:

____

____

____

____

____

Essay:

____

____

____

____

____

____

____

____

____

____

____

____

____

____

____

____

____

____

____

____

____

Should I Get a Flu Shot?

10) Analyze and evaluate YOUR beliefs on getting the annual flu shot.

A) Do YOU believe in getting the flu shot each year? Why or why not?
B) Should governments be allowed to force somebody to get a flu shot? Why or why not? Research both sides of the **argument** and construct a valid argument to defend your belief.
C) After solidifying YOUR belief, construct a valid **counterargument** (why somebody could believe the argument against yours) and defend why they would think that.
D) Once done with that, then re-solidify why YOUR beliefs are correct and why the other side is wrong.

Thesis:

____
____
____
____
____

Counterargument:

____
____
____
____
____

Essay:

____
____
____
____
____
____
____
____
____
____
____
____
____
____
____
____
____
____
____
____

DEBATE – For extra practice debate the merits of getting a flu shot with somebody who agrees with your point of view. Then practice debating the merits of not getting a flu shot with somebody who disagrees with your point of view.

Why does the Polar Vortex Cause Cold Snaps?

11) Examine and report on rising temperatures in the arctic regions of the earth and connect how that is causing colder weather in the upper to mid latitudes during the winter time.

STDs and Teens

12) Support or refute a study by the CDC (Center for Disease Control and Prevention) that over 3 million teenage girls are infected with a sexually transmitted disease (STD). Use the search engine GOOGLE to aid in your research. *Be sure to analyze the factors which can cause teenage girls to become infected with a sexually transmitted disease. Include what can be done to prevent STDs.*

The Cinnamon Challenge: Funny or Dangerous?

13) Analyze the dangers involved in the social phenomenon known as the "Cinnamon Challenge" and report why it has caught on throughout the US and is spreading rapidly through social media even though there are substantial health risks for participating in this viral internet food challenge.

MORE CRASH COURSE BOOKS WRITTEN BY ROGER MORANTEL AND CURRENTLY FOR SALE ON AMAZON INCLUDE:

1) CRASH COURSE PSYCHOLOGY: A STUDY GUIDE OF WORKSHEETS FOR PSYCHOLOGY
2) CRASH COURSE US HISTORY: A STUDY GUIDE OF WORKSHEETS FOR US HISTORY
3) CRASH COURSE GOVERNMENT AND POLITICS: A STUDY GUIDE OF WORKSHEETS FOR GOVERNMENT AND POLITICS
4) CRASH COURSE WORLD HISTORY: A STUDY GUIDE OF WORKSHEETS FOR WORLD HISTORY
5) CRASH COURSE ECONOMICS: A STUDY GUIDE OF WORKSHEETS FOR ECONOMICS
6) CRASH COURSE LITERATURE: A STUDY GUIDE OF WORKSHEETS FOR LITERATURE

*PLEASE EMAIL HOLDEN713@GMAIL.COM TO CONTACT THE EDITOR.

Made in the USA
Middletown, DE
15 April 2022